ELEMENTS OF THE PERIODIC TABLE

# WHAT IS IRON?

KATHLEEN A. KLATTE

Published in 2026 by The Rosen Publishing Group, Inc.
2544 Clinton Street, Buffalo, NY 14224



Portions of this work were originally authored by Heather Hasan and published as *Iron*. All new material this edition authored by Kathleen A. Klatte.

Designer: Rachel Rising
Editor: Kathleen A. Klatte



**Cataloging-in-Publication Data**
Names: Klatte, Kathleen A.
Title: What is iron? / Kathleen A. Klatte.
Description: Buffalo, NY : Rosen Publishing, 2026. | Series: Elements of the Periodic Table | Includes glossary and index.
Identifiers: ISBN 9781499478426 (pbk.) | ISBN 9781499478433 (library bound) | ISBN 9781499478440 (ebook)
Subjects: LCSH: Iron--Juvenile literature. | Transition metals--Juvenile literature. | Chemical elements--Juvenile literature.
Classification: LCC QD181.E4 K538 2026 | DDC 546'.621--dc23

Some of the images in this book illustrate individuals who are models. The depictions do not imply actual situations or events.

*Manufactured in the United States of America*

CPSIA Compliance Information: Batch #CSRYA26. For further information, contact Rosen Publishing at 1-800-237-9932.

# CONTENTS

# INTRODUCTION

Since ancient times, people have looked to the night sky and wondered what's up there. What are those bright lights? What are they made of? Why do they shine? And why do they sometimes fall to Earth?

Today scientists know that everything in the sky is made of the same elements, or building blocks, as everything on Earth. Stars are made of the same hydrogen that surrounds us—just in different proportions.

One of the materials that appears both on Earth and in space is iron. Although today we mine iron ore from the earth, ancient people mostly encountered it as meteorites that fell from the sky. A meteorite is a meteoroid, or piece of space rock, that has passed through the atmosphere (when it's called a meteor) and fallen to Earth. Meteors glow as their outer crust burns due to the friction of passing through the atmosphere.

The pure iron found in the earth tends to rust and disintegrate easily. The iron in meteorites is mixed with small amounts of other metals—usually nickel. This combination of two kinds of metal is called an alloy. Adding nickel to iron makes a stronger metal that's less likely to rust. This is why objects ancient groups, such as the ancient Egyptians, made from iron taken from meteorites still exist today.

Modern metallurgists have discovered different alloys featuring iron to accomplish different tasks. We know that iron, like the other elements, is all around us. Up in the sky, down in the earth—even in the blood that keeps our bodies functioning. Iron is everywhere, so learning more about it is an important part of understanding the world around us.

Meteorites can vary from tiny to extremely large. The Hoba meteorite, made of iron and nickel alloy, was discovered in Namibia in 1920. It's the largest known meteorite found on Earth, estimated at 66.1 tons (60 mt).

# CHAPTER 1
# ELEMENTAL IRON

Iron is the fourth most common element in Earth's crust, of which it forms about 5 percent. It's the second most abundant metal after aluminum. It's also the cheapest and most used. In fact, iron even has a major time period named for it.

The Iron Age lasted from about 1200 BCE to about 550 BCE. This was the time when people learned how to combine metals into alloys, especially steel. Steel is a mixture of iron and carbon. It's much stronger than pure iron. Pure iron mined from the earth rusts and breaks easily and doesn't last long. Iron objects that predate the Iron Age were made from iron meteorites, which are usually alloys containing nickel.

The word "iron" comes from the Anglo-Saxon word *iren*. However, iron's chemical symbol is Fe, from the Latin word *ferrum*.

Iron ore is rock that contains iron. Magnetite and hematite are two of the best-known iron ores.

Because the use of iron dates to ancient civilizations, no one person is credited with its discovery.

## WHAT'S AN ELEMENT?

Everything that humans have encountered so far in the universe is composed of one or more elements. An element is a substance that cannot be broken down into different kinds of atoms by natural means. Each element is made up of only one kind of atom, and each atom of iron is the same. Atoms are very tiny. It would take 200 million average-size atoms, lying side by side, to form a line that is only 0.4 inch (1 cm) long! Amazingly, atoms are made up of even smaller parts—subatomic particles.

Iron is all around us—in the earth, the sun, in manufactured items like cars, and even in your blood!

## SMALLER THAN AN ATOM

There are three major subatomic particles that make up an atom: neutrons, protons, and electrons. Neutrons and protons are clustered together at the center of the atom to form a dense core called the nucleus. Neutrons carry no electrical charge, while protons have a positive electrical charge. This gives the nucleus an overall positive electrical charge. Iron has 26 protons in its nucleus, so its nucleus has a charge of +26.

Around the nucleus of an atom are negatively charged electrons, arranged in layers called shells. The electrons are not fixed in a single position, but orbit, or circle, around the nucleus. Why do the electrons remain in orbit? The negative electrons are attracted to the positive nucleus, and it is this attraction that holds the electrons around the nucleus. So that the positive and negative charges of the atom balance, the number of protons and electrons are almost always equal. Therefore, since iron has 26 protons, it also has 26 electrons.

## IRON SNAPSHOT

**CHEMICAL SYMBOL:** Fe

**PROPERTIES:** Transition metal; silver color; solid at room temperature

**DISCOVERED BY:** Was known by ancient civilizations

**ATOMIC NUMBER:** 26 **ATOMIC WEIGHT:** 55.845

**PROTONS:** 26 **ELECTRONS:** 26 **NEUTRONS:** 30

**DENSITY AT 293 K:** 7.86 g/cm$^3$

**MELTING POINT:** 2,795°F; 1,535°C; 1,808 K

**BOILING POINT:** 4,982°F; 2,750°C; 3,023 K

**COMMONLY FOUND:** Earth's crust as iron ore; meteorites

Pure iron is a silver-colored metal. However, it doesn't stay that way for long. Pure iron reacts with water and oxygen to form rust very easily.

# KEEPING THE ELEMENTS ORGANIZED

Today, we know of more than 100 elements. As more elements were discovered over the years, they had to be organized somehow. Scientists eventually arranged the elements on a big chart called the periodic table of elements. The table appears in classrooms and textbooks all over the world.

Mendeleev theorized that there were elements yet to be discovered that fit into his design for the periodic table. During his lifetime, he was proven correct. Newly discovered elements fit into the spaces he left on the table.

The periodic table that we use today is based on the work of Russian chemist Dmitri Mendeleev (1834–1907), who published the first version of the table in 1869 while teaching chemistry at the University of St. Petersburg in Russia. Mendeleev sought to organize the elements in a way that would make it easier for his students to study and understand them. He arranged the elements in horizontal rows according to atomic weight, with the lightest element of each row on the left and the heaviest on the right. Though Mendeleev's periodic table did not list all the elements that we know of today, iron was among those he included.

Today, the elements on the periodic table are listed in order of increasing atomic number, which is an atom's number of protons. Arranged like this, many trends, or patterns, can be seen. You can use these trends to help you classify the elements. By finding where an element is on the periodic table, you can predict whether it is a metal, a nonmetal, or a metalloid. Metals can be recognized by their physical traits. Generally, metals can be polished to be made shiny. We also know that metals conduct electricity. Most metals also can be hammered into shapes without breaking. This is called malleability. Metals are also usually ductile, which means that they can be pulled into wires. Substances such as plastics, glass, and wood are classified as nonmetals, because they lack the characteristics of metals. Metalloids, or semimetals, have characteristics of both metals and nonmetals. In most respects, metalloids behave like nonmetals. However, they can conduct electricity, though not nearly as well as metals do.

The elements on the periodic table are divided by a "staircase" line. The metals are found to the left of this line and the nonmetals are on the right. Most of the elements bordering the staircase line are metalloids. Iron, as you would expect, is located to the left of the staircase line.

## EVERYTHING IN ITS PLACE

The periodic table is very useful because we can tell a lot about an element just by seeing where it is on the table. As you look across the table from left to right, the horizontal line of elements is called a period. Elements are arranged within periods depending on the number of electron shells that surround the nucleus of their atoms. The outermost electrons determine how elements behave and react with one another. Iron is in period 4 because it has four shells of electrons surrounding its nucleus.

As you read down the chart from top to bottom, the vertical line of elements you see is called a group, or family. Just as you might have similar characteristics to the other members of your family, the elements in a group have similar characteristics called properties. Iron is in group VIIIB, located near the middle of the periodic table. Overall, the elements found in group VIIIB have the properties typical of metals, such as malleability and the ability to conduct electricity.

In addition, some elements are grouped together in triads. Iron, cobalt, and nickel have similar properties, so they have been grouped together to form the iron triad.

Iron is a transition metal. This means it has electrons that can form bonds with other elements. Other important commercial metals such as nickel and titanium are also considered transition metals.

## THE ONLY ONE OF ITS KIND

What makes iron different from other elements like oxygen (O) and silver (Ag)? The difference is the number of protons found in the nucleus of the atoms. Since it is the number of protons that makes each element unique, the elements on the periodic table are organized by these numbers. The number of protons that are found in an atom of an element is called the atomic number. On the periodic table, this number is found above and left of the element's symbol. Since iron has 26 protons, its atomic number is 26. The fact that iron has 26 protons in its nucleus is what makes it iron. If you could add one proton to iron's nucleus—which is physically impossible to do—you would have an entirely different element. By adding another proton, you would have an atom of the element cobalt, which has 27 protons. If you were able to take away one of iron's protons, you would have manganese, which has 25 protons in its nucleus.

# HOW MUCH DOES IRON WEIGH?

The number that is found in the upper-right corner above an element's symbol on the periodic table is called the atomic weight. Iron has an atomic weight of 55.845. When we know the atomic weight of an element, it helps us to figure out how many neutrons there are in an atom of that element. The atomic weight is the approximate sum of the number of protons and neutrons in the atom. Therefore, knowing that the atomic weight of iron is 55.845 and the atomic number (number of protons) is 26, we can figure out how many neutrons there are in an atom of iron by subtracting the atomic number from the atomic weight:

55.845 – 26 = 29.845

Iron has about 30 neutrons in its nucleus. We can find out a lot of information about iron just by looking at the periodic table!

An element's entry on the periodic table contains a lot of information: its atomic number and weight, its chemical symbol, and its common name.

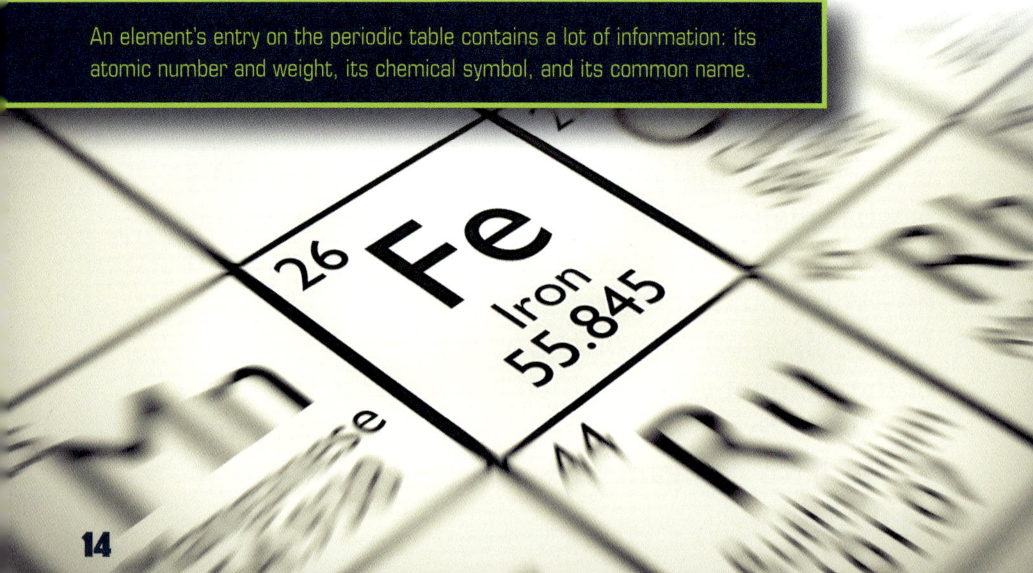

# THE LATEST ADDITIONS

Mendeleev realized that there were probably more elements in the world than had been discovered. He left spaces in the periodic table for new elements to be added. His idea was proven correct, as more elements were discovered and added to the table during his lifetime and beyond. Improvements in technology have enabled the discovery or creation of additional elements. Scientists still make new discoveries today. The latest additions to the periodic table are four superheavy elements created in laboratories. In 2016 elements 113, 115, 117, and 118 were added to the table.

| ATOMIC NUMBER | INITIAL DESIGNATION | OFFICIAL NAME |
|---|---|---|
| 113 | UNUNTRIUM (Uut) | NIHONIUM (Nh) |
| 115 | UNUNPENTIUM (Uup) | MOSCOVIUM (Mc) |
| 117 | UNUNSEPTIUM (Uus) | TENNESSINE (Ts) |
| 118 | UNUNOCTIUM (Uuo) | OGANESSON (Og) |

Shown here are other examples of transitions metals like iron. Like other metals, they can be hammered into different shapes or pulled into wire. They also conduct electricity.

## BRIDGING THE GAP

The transition metals serve as a bridge connecting the two sides of the periodic table. This group includes many common commercially used metals, including iron, titanium, copper, and nickel. It also includes precious metals such as gold, silver, and platinum. The group was recognized and named in 1921 by an English chemist named Charles Bury (1890–1968). Today, the International Union of Pure and Applied Chemistry (IUPAC) defines transition metals as “any element with a partially filled d-electron sub-shell.” IUPAC is the organization that maintains and publishes the current periodic table of elements.

Most elements can only lose or gain a set number of electrons. This means they can only form one kind of ion. An ion is an electrically charged particle. Ions are formed when atoms gain or lose electrons. Generally, atoms have the same number of electrons and protons, so the charges cancel each other, making the atom electrically neutral. An anion has more electrons than protons, meaning it carries a negative charge. A cation has more protons, meaning it carries a positive charge.

Transition metals can form more than one ion. This makes them very useful as catalysts in chemical reactions. Iron can form two ions. $Fe^{2+}$ is a ferrous ion. This means the iron atom has lost two electrons. $Fe^{3+}$ is a ferric ion. This means the iron atom has lost three electrons.

Nickel is a transition metal most people interact with every day. It's used in alloys with iron and steel for construction. It's also used in batteries and in the spark plugs in your car.

# CHAPTER 2

# ALL ABOUT IRON

Just like people, elements have things that make them different from all other elements. These are the element's physical and chemical properties. Physical properties are things that describe how the element interacts with its environment, such as:

- Phase at room temperature (whether the element is a solid, liquid, or gas)
- Hardness
- Melting point
- Color
- Density
- Solubility
- Boiling point

These properties can be observed without changing the nature of the element. For example, oxygen will always be a gas at room temperature.

The chemical properties of an element describe how that element undergoes chemical changes—that is, changes that transform the element into a different substance. For instance, when hard, shiny iron is exposed to hydrogen and oxygen, it transforms into brittle, reddish rust. Unlike strong metal, rust crumbles under minimal pressure. The iron has become a totally different substance.

## PHASE AT ROOM TEMPERATURE

At room temperature, an element is in one of three phases: solid, liquid, or gas. Knowing the phase, or physical state, of an element at room temperature helps scientists identify it. Iron is in the solid phase at room temperature. A solid has a fixed shape and volume. Solids also resist being compressed or having their shape changed.

An element's phase at room temperature is a constant physical property. At room temperature, iron is always solid, mercury is always liquid, and oxygen is always gaseous.

## DENSITY

Iron's density is 7.86 g/cm$^3$. Density is how compact an object is—that is, how much mass it contains per unit volume. Solids are generally denser than liquids, which are, in turn, denser than gases. If you sprinkle iron shavings on water, they will sink. This is because iron is denser than water.

## HARDNESS

When iron is in its pure form, not mixed with anything else, it is a relatively soft metal. On the Mohs' scale, iron has a hardness of 4. For comparison, your fingernail has a hardness of 2.5, and a penny has a hardness of 3.5. Iron is harder than both and could scratch them. However, there are a lot of materials that can scratch pure iron. To make iron harder, other elements are added to it. This is how steel is made.

Steel is an alloy that's harder than pure iron. It also resists rust much better.

# HOW HARD IS IRON?

One of the physical properties of an element is hardness. The scale used to measure hardness is called the Mohs' scale. It was developed in 1812 by German mineralogist Friedrich Mohs (1773–1839). The scale is based on 10 minerals of ascending hardness. Each mineral can scratch the ones that come before it and can be scratched by the ones that come after.

## MOHS' HARDNESS SCALE MINERALS

| MINERAL | HARDNESS |
|---|---|
| TALC | 1 |
| GYPSUM | 2 |
| CALCITE | 3 |
| FLUORITE | 4 |
| APATITE | 5 |
| ORTHOCLASE | 6 |
| QUARTZ | 7 |
| TOPAZ | 8 |
| CORUNDUM | 9 |
| DIAMOND | 10 |

continues on the next page

On the Mohs' scale, talc (1) is the softest mineral and diamond (10) is the hardest. Everything else on the scale is harder than talc and can scratch it. Diamond can scratch everything else on the list. There are pocket Mohs' kits available for use in the field. They tend not to include diamond because it would be too expensive!

Some everyday objects can also be compared to things on the Mohs' scale. A human fingernail is about a 2.5 and can scratch talc and gypsum. A masonry drill bit is about an 8.5 and can scratch topaz.

## CONDUCTION

Like all metals, iron conducts electricity and heat. This means that heat and electrical current can move through iron. Metals can conduct electricity because the electrons from the outer shells of their atoms (valence electrons) are able to move about in what is called a "sea" of electrons. This movement of electrons is an electric charge, or electricity. The sea of electrons that metals have also makes them good conductors of heat. When metals such as iron are heated, the electrons gain more energy. This makes them move about more quickly, distributing the heat throughout the whole metal.

It takes a lot of heat to pull apart iron atoms. Iron's melting point is 2,795°F (1,535°C). Because solid iron can handle a lot of heat without melting, it is often used to build things that are exposed to intense heat, such as a car engine.

Because it withstands heat so well, cast iron has been used for centuries to make cooking pots

Once iron is melted into a liquid, the temperature must reach a steamy 5,432°F (3,000°C) before iron will boil and become a gas. To give you an idea of how hot that is, you can compare this temperature to the temperature found on the surface of the sun, about 10,000°F (5,538°C). Iron would definitely boil there!

Magnetite is an iron oxide mineral. As you can see, it's quite magnetic!

## MAGNETISM

One of iron's most important properties is its strong attraction to magnets. Nickel and cobalt are also magnetic, but iron is much more common than these elements. In fact, if you want to know if an object contains iron, you can test it with a magnet. Magnetism is a force that pulls things together or forces them apart.

The word "magnetism" comes from the ancient Greek region of Magnesia, a place where lodestones were found in ancient times. Lodestones are rocks that contain iron and were the first known magnets. Although these rocks had become permanently magnetized, most iron is not permanently magnetic. The fact that iron can become magnetic is due to the structure of the iron atom. The electrons that orbit the nucleus of the iron atom make each atom act like a tiny magnet. If the electrons are lined up in just the right way, a piece of iron becomes magnetized. Iron is made magnetic when it is properly exposed to a magnetic field. A magnetic field can be produced by another magnet or by an electric current. Heating or melting the iron, however, can make the metal lose its magnetism.

These are the remains of a temple to Artemis in Magnesia. The temple is believed to have been built between 200 and 130 BCE. Traditionally, a Greek philosopher named Thales (circa 600 BCE) is believed to be one of the first people to write about magnetism.

The aurora borealis or northern lights are the result of a magnetic storm caused by solar flares.

You can think of Earth as a huge magnet. Earth's core is made mostly of iron. The core is so hot that the outer part of the core is always liquid. Scientists think that, as Earth spins, the liquid iron swirls around. This movement creates Earth's magnetism.

Earth has magnetic poles, just like a bar magnet. Earth's magnetism is the strongest at these poles. If a small magnet is allowed to turn freely, it will point toward earth's poles. That is how a compass works—the magnet inside points toward the North Pole and you can use that to find your direction.

# I SPY SOME IRON

Most of the iron ore in the earth is found in sedimentary rock deposits. These rocks formed about 1.8 billion years ago. Iron atoms bond easily with other elements, so it's extremely rare to find pure iron outside a lab. Most iron is combined with other elements to form different minerals. For instance, iron is commonly found in iron oxides such as hematite and magnetite. These minerals formed centuries ago on the ocean floor when the oxygen in water combined with iron.

Pure iron is a silvery, shiny metal. When iron atoms bond with other elements, a chemical change occurs. The resulting minerals have different colors and textures. Hematite can be reddish, and magnetite looks something like coal.

Iron ore is often found in sedimentary rocks—that is, rocks formed of different layers of materials compressed together. These rocks may be different colors, depending on what other minerals are part of the rock. They may also be banded.

# CHAPTER 3

# ORE, STEEL, AND HEMOGLOBIN

Iron is found throughout the universe. It's the most common metal in the universe and the second most common metal on earth. Iron is found in the earth's core, inside stars, and even inside you! Iron makes up about 5 percent of the earth's crust. It may also be part of meteorites that have fallen to Earth. However, pure iron doesn't occur in nature. Regardless of whether it's mined from the earth or fallen from space, iron is always bonded with other elements. Pure iron is obtained by a process called smelting. This involves heating ore to the melting point and using oxidizing agents or reducing agents to extract the metal from rock.

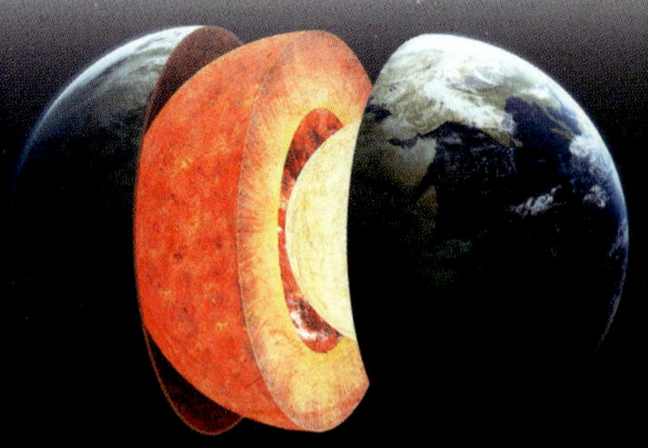

The iron found in the earth's crust isn't pure—it's part of rocks containing other elements.

## GOT ROCKS?

Iron ores are rocks that contain iron and other elements. Iron was concentrated in these rocks by natural forces during the formation of Earth's crust millions of years ago. People have been trying to extract iron from these rocks for thousands of years, and it is from these ores that we get the iron we use today. Our ability to get iron from ore has made iron a very commonly used metal today. Because iron is so common, it is one of the world's cheapest but most useful metals.

Lava (melted rock) from a volcanic eruption reaches temperatures of 1,300 to 2,200°F (700 to 1,200°C). That's much less than the melting point of iron.

The main ores from which we get iron are hematite, magnetite, and pyrite. People mine these iron ores. Of these iron ores, hematite and magnetite are the richest in iron, containing about 70 percent of the metal by weight. These ores contain iron and oxygen and are called iron oxides. Hematite may be black, brownish red, or dark red. Magnetite is black and has magnetic properties. Pyrite is about half iron and half sulfur (S). You may have heard of pyrite, which is commonly called fool's gold. Its shiny, gold-colored appearance has fooled many people into thinking that they had found gold! Iron is also extracted from limonite, siderite, and taconite ores.

## THREE KINDS OF IRON

There are three major types of iron alloys: cast iron, wrought iron, and pig iron. Cast iron is an iron alloy and contains iron and two other elements, carbon and silicon (Si). Because it contains carbon, cast iron is very hard. However, cast iron is also very brittle, which means it breaks apart easily. It cannot be shaped with a hammer even if it is heated to a very high temperature; it will simply break. Cast iron is made into useful things when it is melted and then allowed to cool in a mold. The liquid iron takes the shape of the mold, and when it cools and hardens, it keeps that shape. You have seen cast iron in manhole covers, automobile parts, and cookware.

Wrought iron is nearly pure iron. It is mixed with just a tiny bit of a glasslike material called iron silicate. Wrought iron resists corrosion, or rusting, better than cast iron does. Unlike cast iron, wrought iron is malleable. This means that it can be hammered into different shapes. Because it is malleable, wrought iron is used to make metal fences that have a lot of fancy designs. You have also seen wrought iron in coatracks and in the handrails on staircases.

Pig iron is made in a blast furnace. It contains some carbon and small amounts of other elements. At one time, people shaped liquid pig iron by taking it from a blast furnace and pouring it into molds. Pig iron got its name because the molds that it was poured into looked a little like a group of baby pigs lined up around a mother pig. Today, most pig iron is used to make steel.

A blast furnace is much hotter than the furnace that heats your home. A blast furnace can reach temperatures of 3,000°F (1,650°C).

Steel is everywhere in the modern world. British engineer Henry Bessemer (1813–1898) is credited with inventing an economical process for manufacturing steel.

## STEEL

Most elemental iron is used to make steel. You can think of steel as refined, or purified, iron. Unlike pure iron, steel is not an element. Steel is produced by refining iron and then mixing it with other elements. Steelmaking involves first removing unwanted substances, such as excess carbon, silicon, or sulfur. To do this, the iron ore is mixed with limestone and heated to a very high temperature. This is normally done in one of three furnaces: an open hearth, an electric furnace, or a basic oxygen furnace. Once this is done, the desired materials are carefully added. These added materials make steel a lot stronger than iron.

# FOUR KINDS OF STEEL

There are thousands of kinds of steel, but these can be grouped into four types: carbon steel, alloy steel, stainless steel, and tool steel. Carbon steel is used more often than any other kind of steel. The properties of this steel depend on how much carbon it contains. The higher the percentage of carbon, the harder the steel is. Carbon steel is made into many products, such as structural beams in buildings, automobile bodies, kitchen appliances, and cans.

Alloy steel contains some carbon, but its properties depend on the other elements it contains. Each element that is added to alloy steel improves one or more of its properties. Nickel, for example, makes the steel stronger. When manganese is added, it makes the steel harder, tougher, and more resistant to wear. Molybdenum, another element, increases the steel's hardness and resistance to corrosion. Other elements used in alloy steel include chromium, aluminum, tungsten, copper, titanium, silicon, and vanadium.

A wide variety of manufactured goods are made of steel. The tools and equipment used to make those goods are often also made of steel.

Stainless steel contains a large amount of chromium. Many stainless steels also contain nickel. This type of steel resists corrosion better than any other steel. It is used to make kitchen items such as knives, flatware, pots, and pans. You can also find stainless steel in things such as automobile parts, hospital equipment, and razor blades.

Tool steel is a very hard steel. It is made by a process called tempering. In this process, certain types of carbon steel and alloy steel are heated to a high temperature and then cooled very quickly. Because it is so hard, tool steel is used in metalworking tools such as files, drills, and chisels. Tools made from tool steel are then used to cut other, softer metals.

## IRON INSIDE YOU

Iron is all around us—in trains and cars, the rails and bridges they run on, and the garages that house them. There's iron in the pipes that carry water to your house and the wires that carry electricity. There's also iron inside you.

Your body contains about 0.005 percent iron. Most of that iron—about 65 percent—is in a substance called hemoglobin. Hemoglobin is a protein in your blood that carries oxygen throughout your body. Hemoglobin—and the iron it contains—is why your blood is red.

People need a certain amount of iron a day to stay healthy. But where does it come from? After all, we don't eat nails and screws for breakfast! Iron comes from eating foods like meat, eggs, bread, fruit and vegetables. Besides the hemoglobin currently circulating in your bloodstream, your body stores iron in your liver, spleen, and bone marrow to make more hemoglobin when it needs it.

Eating a balanced diet is the best way to ensure you're getting the proper amount of iron to keep your body healthy—not too much and not too little.

Not having enough iron in your body can make you sick. This is why it's important to eat foods rich in iron. Eating fruits rich in vitamin C helps your body to process the iron in food. Without enough iron, you can feel tired and run down. Iron-deficiency anemia is the medical name for not having enough iron in your body. Sometimes people don't have enough iron in their blood because they don't eat enough iron-rich foods. There are also illnesses that affect how the body produces red blood cells.

## TOO MUCH OF A GOOD THING?

Your body needs iron to be healthy. Generally, if you eat a balanced diet, you should get enough iron. However, certain groups of people may need more than others. Pregnant people often require extra iron because their body is using lots of nutrients and blood growing a baby. Vegetarians are also at risk because a human body absorbs iron from meat more readily than from plants.

Your body needs a certain amount of iron to function, but too much can make you sick. Some people suffer from a condition called iron overload, which happens when their bodies store too much iron. One cause of this is a hereditary disease called hemochromatosis.

Vegetables, red meat, and legumes are good sources of iron.

# CHAPTER 4

# MIXING IT UP!

Most of the elements of the periodic table can bond with other elements to form compounds. When a metal like iron bonds with a nonmetal, it forms an ionic compound. This means that electrons have transferred from the metal to the nonmetal. Ionic compounds contain positive and negative ions. Because iron can share either two or three electrons, it can form multiple compounds with the same other element.

- Iron that's lost two electrons ($Fe^{2+}$) forms ferrous compounds.
- Iron that's lost three electrons ($Fe^{3+}$) forms ferric compounds.

For example, iron can form two different compounds with oxygen (iron oxides): ferrous oxide and ferric oxide. Ferrous oxide, also called wustite, is a black powder used in manufacturing steel, enamel, and certain types of heat-absorbing glass. Ferric oxide, or hematite, is red. It's used in the iron and steel industries.

While some forms of iron oxide are useful, the form most people are familiar with is neither useful nor welcome—rust. Rust is the result of a chemical reaction between iron, oxygen, and water.

## GETTING RUSTY

We have probably all had experience with rust. If you have ever left your bicycle outside in the rain, you might have experienced it firsthand! The reddish-brown or reddish-yellow powder that you found on your bicycle when you finally brought it inside is rust. Rusting requires a metal, oxygen, and water. Iron will not rust in pure water that is free of extra oxygen, and it will not rust in pure oxygen that is moisture free.

Rust isn't just ugly—it breaks down the structure of metal. Instead of being hard and shiny, this metal is weak and crumbly. The components of the car that the metal is intended to protect are now open to the weather.

When iron, oxygen, and water meet, rust is readily formed. Rust is the common name for ferric oxide that's formed by this chemical change. It has the chemical formula $Fe_2O_3$. Ferric oxide that occurs naturally is called hematite. It's an ore containing iron which can occur in different forms. The characteristic red color of rust can also be found in hematite.

People do many things to keep their iron and steel possessions from rusting. One way to protect these metals is to paint them. The paint on cars forms a barrier between the iron that the car is made of and the water and oxygen in the air. Likewise, many household items have plastic coatings to protect them. A lot of storage shelves are wrought iron covered with plastic. Another way to protect iron and steel is to plate, or cover, it with a metal that does not rust, such as chromium (Cr). Car bumpers are sometimes made of chrome-plated steel.

There are entire industries devoted to rust-proofing metal. For plumbing, the aim is to preserve the integrity of the pipes, both to prevent damage to the building and to prevent harmful minerals from leeching into drinking water from damaged metal.

# THE RED PLANET

When ancient people looked up to the night sky and saw a red planet, they named it for its color. The Egyptians called it Her Dasher, meaning "red one." The Romans called it Mars, for their god of war, because the red reminded them of blood.

Today, scientists have landed vehicles on Mars to study the surface. They know that the red color comes from all the iron on the planet's surface. Scientists now believe that billions of years ago, Mars had a thicker atmosphere, with water and oxygen that oxidized the iron in the soil, turning it to rust.

The gravity on Mars isn't as strong as Earth's. Scientists think this might be a reason why so much iron remains on the surface of the planet, instead of being concentrated at the core, like it is on Earth.

Because rust is the product of a chemical change, it cannot be separated back into iron and oxygen by any physical means, such as boiling or freezing. Recovering the iron and the oxygen from rust would require another chemical reaction. One way to get rid of rust is by "pickling" the iron. In this process, the rust reacts with an acid. This turns the rust into a harmless substance that can be washed away from the metal. Sometimes hydrochloric acid (HCl) is used for this. This acid reacts with the rust to form a useful iron compound called ferric chloride ($FeCl_3$).

## USING IRON FOR HEALTHY WATER

Ferric chloride is a dark-colored crystal. This compound is used mainly for water treatment. It helps to remove impurities from water by causing them to lump together. Once the unwanted material has formed a mass, it can then be filtered out of the water.

Ferric chloride helps to remove soil and other unwanted substances from our drinking water. It is also used by the food, paper, photography, and pharmaceutical industries. All these industries need to use very pure water. Ferric chloride is also helpful for the environment. Before wastewater is released into our rivers and lakes, it is treated with ferric chloride. This makes it safe enough so that it will not harm the organisms that live in the water.

Ferric chloride is also sometimes used as a disinfectant to destroy harmful bacteria, viruses, and other disease-causing materials. These things are not only harmful to us, but they also give water a bad taste, an odor, or a funny color. Ferric chloride treatment helps to make drinking water look and taste better, and it also helps sewage smell better!

## COLORFUL IRON

Ferrous chloride ($FeCl_2$) is formed when hot chlorine gas is passed over iron. This is a ferrous compound because iron has donated two electrons to the chlorine. Ferrous chloride is a colorless, crystalline substance. Like ferric chloride, it is used for sewage treatment. However, ferrous chloride is also used for dyeing fabric.

Ferrous chloride is what is known as a mordant. Mordants keep the color from fading from fabric. When dyeing something with a natural dye, which is obtained from plants or animals, it is necessary to use a mordant. If a mordant isn't used, the color will fade when the fabric is exposed to sunlight or washed. Ferrous chloride binds these natural dyes to the fabric, holding them there tightly so that they cannot be washed out.

Another iron compound is ferrous sulfate. Its chemical makeup is shown here (a combination of iron, oxygen, and sulfur), along with its physical appearance.

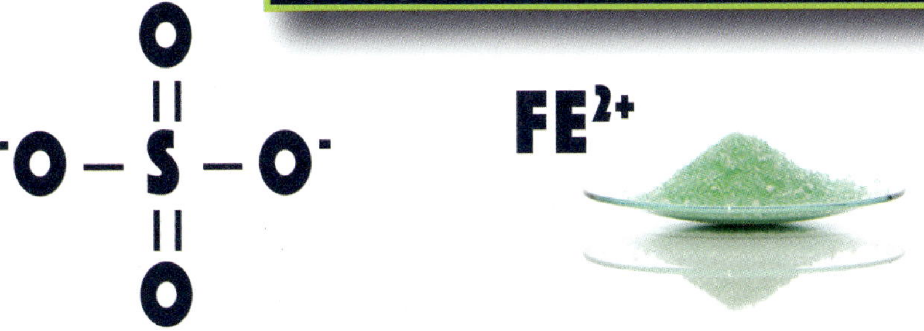

# USING IRON FOR HEALTHY BLOOD

Ferrous sulfate is green, with a crystalline structure. It's an iron compound used to treat iron-deficiency anemia. People who suffer from anemia can feel weak and get tired easily. They are also often cold. Sometimes people develop anemia because they don't get enough iron in their diet. Doctors can remedy this by prescribing iron supplements, usually in the form of ferrous sulfate.

Certain groups of people are more likely to need iron supplements. Pregnant women, menstruating women, and teens often need extra iron.

Your body stores some of the iron you ingest in food in a blood protein called ferritin. If a doctor suspects you might have anemia, they will perform blood tests. One of these might be to check the amount of ferritin in your blood.

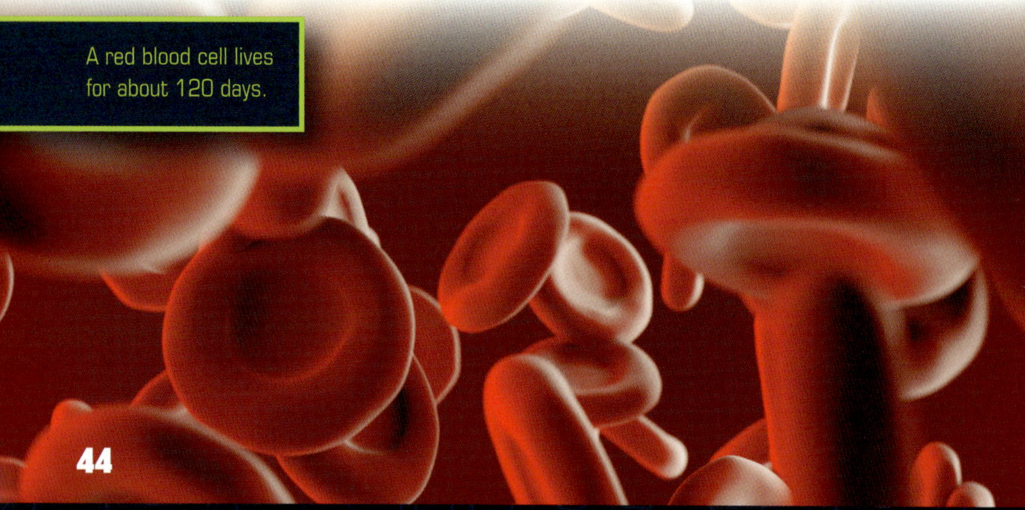

A red blood cell lives for about 120 days.

Red blood cells are produced in your bone marrow. This is the soft tissue found in the center of your bones.

# CHAPTER 5

# IRON IN EVERYDAY LIFE

Your body needs iron to work properly. For most people, the best way to get the iron they need is to eat a balanced diet. Many kinds of food contain iron. However, it's easier for your body to process the iron contained in animal proteins, such as meat and fish. Fruits, nuts, and vegetables also contain iron, but it can be harder for your body to process iron from plants.

In addition to eating food rich in iron, it's important to eat fruits and vegetables that contain vitamin C. Vitamin C helps your body absorb the iron in your food. There are also foods that make it harder for your body to absorb iron. These include coffee, tea, and dairy products.

Some kinds of food have additional iron added to them during processing. These include bread, cereal, and some juices.

Lots of people don't like to eat things like liver or greens. Adding extra iron to foods like bread and cereal makes it easier for people to get the iron they need.

## IRON IN YOUR CEREAL

You can separate tiny pieces of iron from your cereal with just a few materials. You will need:

- a plastic bottle with a screw cap
- warm water
- a large magnet
- iron-fortified cereal (one that lists iron in the nutritional label).

Fill your bottle about one third with water, then add some pieces of cereal. Screw the lid tightly, then shake. If your cereal doesn't break up, you may need to let it soak overnight. Once your cereal has broken up in the water, press your magnet to the outside of the bottle. The iron filings will collect on the inside of the bottle near the magnet! The iron will look like small dark dots. If you have any trouble seeing the iron, try using a magnifying glass.

## HOW MUCH IRON DO YOU NEED?

People need different amounts of iron in their diets depending upon how old they are. Toddlers (ages one to three years old) need at least 3.7 milligrams of iron per day. Young children (ages four to six years) require at least 3.3 mg every day. Older children (ages seven to 10 years) should eat at least 4.7 mg of iron a day. Teenagers (ages 11 to 18) need more iron than any other age group. Teenage girls require at least 8.0 mg of iron, while teenage boys need 6.1 mg.

## IRON CONTENT OF COMMON FOODS

| FOOD | SERVING SIZE | AMOUNT OF IRON |
|---|---|---|
| SKIRT STEAK | 6 OZ | 9.3 MG |
| WHITE BEANS | 1 CUP | 6.6 MG |
| SPINACH | 1 CUP (COOKED) | 6.4 MG |
| PUMPKIN SEEDS | 1 OZ | 2.5 MG |
| FORTIFIED CEREAL | ¾ CUP | 19.6 MG |
| UNSWEETENED BAKING CHOCOLATE | 1 OZ | 5 MG |

## IT'S MAGNETIC!

We know that magnets are attracted to iron, steel, and some other metals, but are they attracted to money? Find a magnet and gather some money to see.

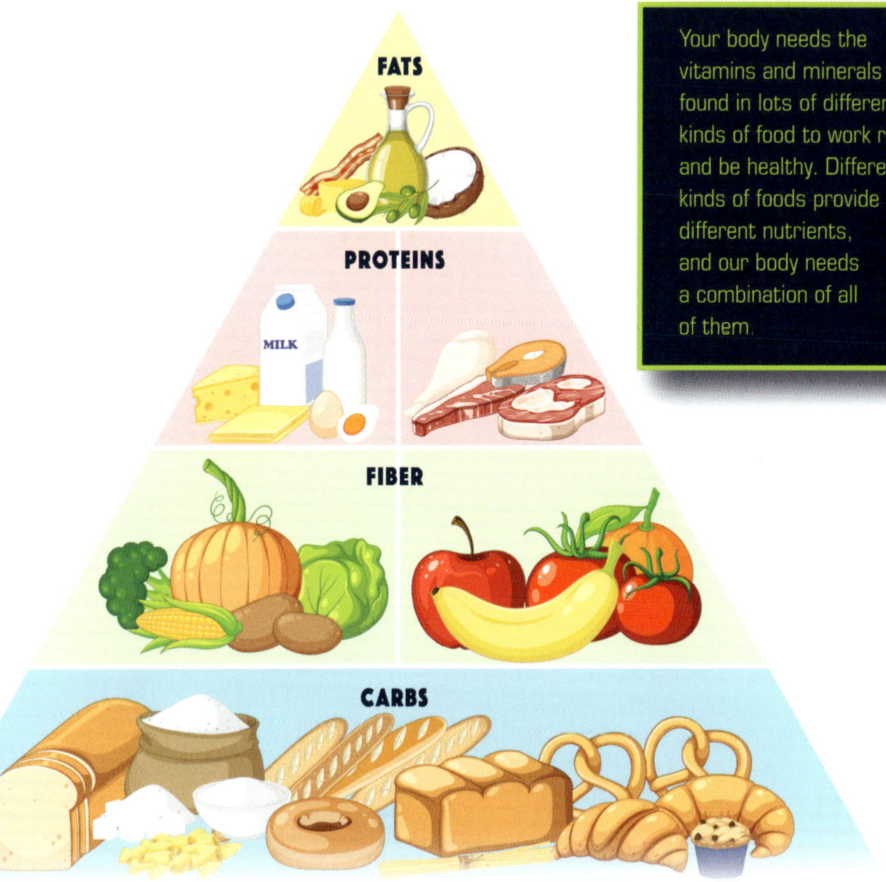


Your body needs the vitamins and minerals found in lots of different kinds of food to work right and be healthy. Different kinds of foods provide different nutrients, and our body needs a combination of all of them.

In the United States, the metal used to make pennies has changed over time. Currently, pennies are made of copper and zinc. Neither of these metals are magnetic, so pennies would not be attracted to the magnet. However, in 1943, pennies were made of steel, so these pennies would be attracted to the magnet. If you find one of these pennies, hold on to it. It could be a collector's item!

You will find that none of the other U.S. coins are attracted to magnets, not even nickels. Even though nickel is a magnetic metal, just like iron, nickel coins do not contain enough of the metal to make them magnetic.

Steel pennies were manufactured during World War II. Before that, pennies were made of copper, zinc, and tin. All those materials were needed for the war effort, so the government looked for alternative metals.

## FINDING YOUR WAY

A compass is a device that you can use to help you find your way. At the center of a compass is a needle made of magnetized iron. No matter where you are, the needle on a compass always points north, toward the magnetic North Pole. Although we now have more complex methods of finding our way, such as maps and GPS (global positioning systems), people have always relied on compasses to get around.

# HERE, THERE, AND EVERYWHERE

Iron is all around us. It's part of the natural world. Iron exists as ore deep in the rocks that form the earth's crust and in the planet's molten core. It's in the sun and all the other stars in the sky. There's iron in the plants that grow and the animals that roam the earth. There's iron in the food you eat and even inside your own body.

Iron also exists in man-made objects. Iron, usually in the form of steel, makes up cars, planes, buses, and trains. Large buildings like schools and hospitals are built around frames of heavy steel beams. There's iron on the playground, too—in the chain link fence that keeps balls from rolling out into the street and in the slides and swings.

There's iron in small things, too. Go for a walk with a magnet and see what you can find. Check out things like nails, screws, coins, and paperclips. Although there are other metals that are magnetic, it's a good bet that most of what you encounter will be at least partly made of iron.

Have you ever walked past a construction site? Those heavy beams that form the skeleton of a building are made of steel.

# MAKE YOUR OWN COMPASS

Are you interested in how a compass works? Well, you're not alone. Albert Einstein (1879–1955), one of the most famous scientists who ever lived, was given a compass as a small boy. Would you like to make your own simple compass?

**YOU'LL NEED:**

- a large needle
- a cork
- a bowl of water
- a magnet

First, you will have to magnetize the needle. Stroke the needle with the magnet about 50 times. The direction should be from the back of the needle to the point. This will make the needle magnetic. Next, flip your needle and repeat the process with the reverse side of the magnet. Push the needle through the cork. Place the cork in the bowl of water. The needle and cork will slowly turn so that the pointed end of the needle points north. Now you have your very own compass! Try turning the needle around and see what happens.

Long before GPS, people used compasses to find their way around.

# THE PERIODIC TABLE OF ELEMENTS

Studying the periodic table can teach us a lot about the elements, including iron. What can you learn from looking at its entry?

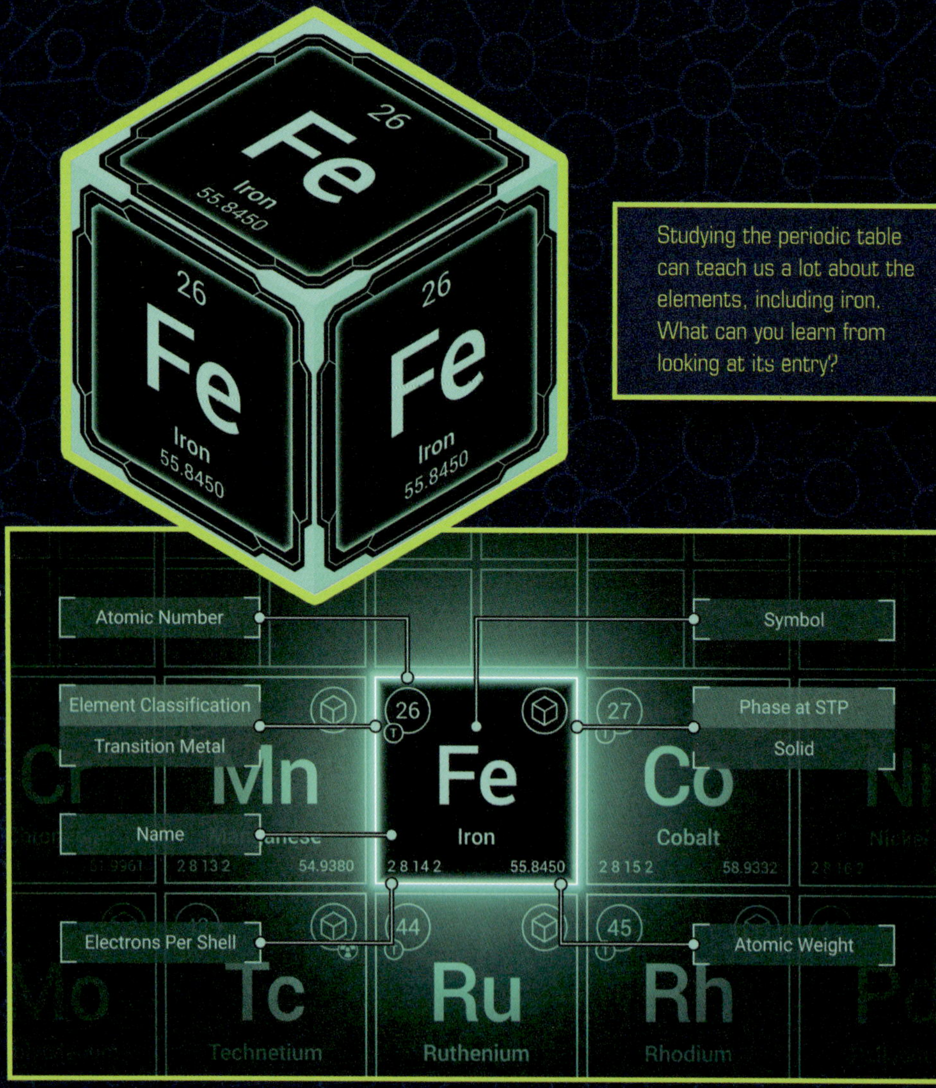

## PERIODIC TABLE OF ELEMENTS

- Alkali metals
- Alkaline earth metals
- Transition metals
- Metal
- Metalloids
- Nonmetals
- Halogen
- Noble gases
- Lanthanoids
- Actinoids

| 1 H Hydrogen 1.008 | | | | | | | | | | | | | | | | | 2 He Helium 4.0026 |
|---|---|---|---|---|---|---|---|---|---|---|---|---|---|---|---|---|---|
| 3 Li Lithium 6.94 | 4 Be Beryllium 9.0122 | | | | | | | | | | | 5 B Boron 10.81 | 6 C Carbon 12.011 | 7 N Nitrogen 14.007 | 8 O Oxygen 15.999 | 9 F Fluorine 18.998 | 10 Ne Neon 20.180 |
| 11 Na Sodium 22.990 | 12 Mg Magnesium 24.305 | | | | | | | | | | | 13 Al Aluminium 26.982 | 14 Si Silicon 28.085 | 15 P Phosphorus 30.974 | 16 S Sulfur 32.06 | 17 Cl Chlorine 35.45 | 18 Ar Argon 39.948 |
| 19 K Potassium 39.098 | 20 Ca Calcium 40.078 | 21 Sc Scandium 44.956 | 22 Ti Titanium 47.867 | 23 V Vanadium 50.942 | 24 Cr Chromium 51.996 | 25 Mn Manganese 54.938 | 26 Fe Iron 55.845 | 27 Co Cobalt 58.933 | 28 Ni Nickel 58.693 | 29 Cu Copper 63.546 | 30 Zn Zinc 65.38 | 31 Ga Gallium 69.723 | 32 Ge Germanium 72.630 | 33 As Arsenic 74.922 | 34 Se Selenium 78.971 | 35 Br Bromine 79.904 | 36 Kr Krypton 83.798 |
| 37 Rb Rubidium 85.468 | 38 Sr Strontium 87.62 | 39 Y Yttrium 88.906 | 40 Zr Zirconium 91.224 | 41 Nb Niobium 92.906 | 42 Mo Molybdenum 95.95 | 43 Tc Technetium (98) | 44 Ru Ruthenium 101.07 | 45 Rh Rhodium 102.91 | 46 Pd Palladium 106.42 | 47 Ag Silver 107.87 | 48 Cd Cadmium 112.41 | 49 In Indium 114.82 | 50 Sn Tin 118.71 | 51 Sb Antimony 121.76 | 52 Te Tellurium 127.60 | 53 I Iodine 126.90 | 54 Xe Xenon 131.29 |
| 55 Cs Caesium 132.91 | 56 Ba Barium 137.33 | 57-71 | 72 Hf Hafnium 178.49 | 73 Ta Tantalum 180.95 | 74 W Tungsten 183.84 | 75 Re Rhenium 186.21 | 76 Os Osmium 190.23 | 77 Ir Iridium 192.22 | 78 Pt Platinum 195.08 | 79 Au Gold 196.97 | 80 Hg Mercury 200.59 | 81 Tl Thallium 204.38 | 82 Pb Lead 207.2 | 83 Bi Bismuth 208.98 | 84 Po Polonium (209) | 85 At Astatine (210) | 86 Rn Radon (222) |
| 87 Fr Francium (223) | 88 Ra Radium (226) | 89-103 | 104 Rf Rutherfordium (267) | 105 Db Dubnium (268) | 106 Sg Seaborgium (269) | 107 Bh Bohrium (270) | 108 Hs Hassium (277) | 109 Mt Meitnerium (278) | 110 Ds Darmstadtium (281) | 111 Rg Roentgenium (282) | 112 Cn Copernicium (285) | 113 Nh Nihonium (286) | 114 Fl Flerovium (289) | 115 Mc Moscovium (290) | 116 Lv Livermorium (293) | 117 Ts Tennessine (294) | 118 Og Oganesson (294) |

| 57 La Lanthanum 138.91 | 58 Ce Cerium 140.12 | 59 Pr Praseodymium 140.91 | 60 Nd Neodymium 144.24 | 61 Pm Promethium (145) | 62 Sm Samarium 150.36 | 63 Eu Europium 151.96 | 64 Gd Gadolinium 157.25 | 65 Tb Terbium 158.93 | 66 Dy Dysprosium 162.50 | 67 Ho Holmium 164.93 | 68 Er Erbium 167.26 | 69 Tm Thulium 168.93 | 70 Yb Ytterbium 173.05 | 71 Lu Lutetium 174.97 |
|---|---|---|---|---|---|---|---|---|---|---|---|---|---|---|
| 89 Ac Actinium (227) | 90 Th Thorium 232.04 | 91 Pa Protactinium 231.04 | 92 U Uranium 238.03 | 93 Np Neptunium (237) | 94 Pu Plutonium (244) | 95 Am Americium (243) | 96 Cm Curium (247) | 97 Bk Berkelium (247) | 98 Cf Californium (251) | 99 Es Einsteinium (252) | 100 Fm Fermium (257) | 101 Md Mendelevium (258) | 102 No Nobelium (259) | 103 Lr Lawrencium (266) |

- C Solid
- Hg Liquid
- H Gas
- Rf Unknown

# ABOUT THE AUTHOR

**KATHLEEN A. KLATTE** is the author of many nonfiction books for children and teens. Topics she has written about range from animals and nature to constitutional law to unusual career choices. She lives in New York with one cat and far too many books and Legos.

# PHOTO CREDITS

Cover, p. 1 remotevfx.com/Shutterstock.com; Cover, pp. 1, 3-64 pluie_r/Shutterstock.com; Cover, p. 1, 3-64 Oksancia/Shutterstock.com; p. 5 marzia franceschini/Shutterstock.com; p. 6 foto_and_video/Shutterstock.com; p. 7 Dzuba Pavel/Shutterstock.com; pp. 8, 16 Peter Hermes Furian/Shutterstock.com; p. 9 albedomusic/Shutterstock.com; p. 10 Adriana Iacob/Shutterstock.com; p. 13 Adisak Riwkratok/Shutterstock.com; p. 14 Antoine2K/Shutterstock.com; p. 17 Tricky_Shark/Shutterstock.com; p. 19 RHJPhtotos/Shutterstock.com; p. 20 InWay/Shutterstock.com; p. 23 EJ Grubbs/Shutterstock.com; p. 24 Michael LaMonica/Shutterstock.com; p. 25 Selcuk Oner/Shutterstock.com; p. 26 Denis Belitsky/Shutterstock.com; p. 27 Chem_Geo_Photo/Shutterstock.com; p. 28 Vadim Sadovski/Shutterstock.com; p. 29 ImageBank4u/Shutterstock.com; p. 31 Nordroden/Shutterstock.com; p. 32 vyskoczilova/Shutterstock.com; p. 33 Nuttawut Uttamaharad/Shutterstock.com; pp. 35, 37 Tatjana Baibakova/Shutterstock.com; p. 39 Yankovsky88/Shutterstock.com; p. 40 Woodpond/Shutterstock.com; p. 41 DIAMOND HEART/Shutterstock.com; p. 43 Chadchai Krisadapong/Shutterstock.com; p. 44 Vladimir Zotov/Shutterstock.com; p. 45 Ground Picture/Shutterstock.com; p. 46 YARUNIV Studio/Shutterstock.com; p. 49 GraphicsRF.com/Shutterstock.com; p. 50 Alysson M/Shutterstock.com; p. 51 zhengzaishuru/Shutterstock.com; p. 52 Shutterstock AI/Shutterstock.com; p. 53 Yankovsky88/Shutterstock.com; p. 54 ShannonChocolate/Shutterstock.com; p. 55 View_From_My_Lens/Shutterstock.com.

## PERIODIC TABLE OF ELEMENTS

- Alkali metals
- Alkaline earth metals
- Transition metals
- Metal
- Metalloids
- Nonmetals
- Halogen
- Noble gases
- Lanthanoids
- Actinoids

| 1 | 2 | 3 | 4 | 5 | 6 | 7 | 8 | 9 | 10 | 11 | 12 | 13 | 14 | 15 | 16 | 17 | 18 |
|---|---|---|---|---|---|---|---|---|---|---|---|---|---|---|---|---|---|
| 1 H Hydrogen 1.008 | | | | | | | | | | | | | | | | | 2 He Helium 4.0026 |
| 3 Li Lithium 6.94 | 4 Be Beryllium 9.0122 | | | | | | | | | | | 5 B Boron 10.81 | 6 C Carbon 12.011 | 7 N Nitrogen 14.007 | 8 O Oxygen 15.999 | 9 F Fluorine 18.998 | 10 Ne Neon 20.180 |
| 11 Na Sodium 22.990 | 12 Mg Magnesium 24.305 | | | | | | | | | | | 13 Al Aluminium 26.982 | 14 Si Silicon 28.085 | 15 P Phosphorus 30.974 | 16 S Sulfur 32.06 | 17 Cl Chlorine 35.45 | 18 Ar Argon 39.948 |
| 19 K Potassium 39.098 | 20 Ca Calcium 40.078 | 21 Sc Scandium 44.956 | 22 Ti Titanium 47.867 | 23 V Vanadium 50.942 | 24 Cr Chromium 51.996 | 25 Mn Manganese 54.938 | 26 Fe Iron 55.845 | 27 Co Cobalt 58.933 | 28 Ni Nickel 58.693 | 29 Cu Copper 63.546 | 30 Zn Zinc 65.38 | 31 Ga Gallium 69.723 | 32 Ge Germanium 72.630 | 33 As Arsenic 74.922 | 34 Se Selenium 78.971 | 35 Br Bromine 79.904 | 36 Kr Krypton 83.798 |
| 37 Rb Rubidium 85.468 | 38 Sr Strontium 87.62 | 39 Y Yttrium 88.906 | 40 Zr Zirconium 91.224 | 41 Nb Niobium 92.906 | 42 Mo Molybdenum 95.95 | 43 Tc Technetium (98) | 44 Ru Ruthenium 101.07 | 45 Rh Rhodium 102.91 | 46 Pd Palladium 106.42 | 47 Ag Silver 107.87 | 48 Cd Cadmium 112.41 | 49 In Indium 114.82 | 50 Sn Tin 118.71 | 51 Sb Antimony 121.76 | 52 Te Tellurium 127.60 | 53 I Iodine 126.90 | 54 Xe Xenon 131.29 |
| 55 Cs Caesium 132.91 | 56 Ba Barium 137.33 | 57-71 | 72 Hf Hafnium 178.49 | 73 Ta Tantalum 180.95 | 74 W Tungsten 183.84 | 75 Re Rhenium 186.21 | 76 Os Osmium 190.23 | 77 Ir Iridium 192.22 | 78 Pt Platinum 195.08 | 79 Au Gold 196.97 | 80 Hg Mercury 200.59 | 81 Tl Thallium 204.38 | 82 Pb Lead 207.2 | 83 Bi Bismuth 208.98 | 84 Po Polonium (209) | 85 At Astatine (210) | 86 Rn Radon (222) |
| 87 Fr Francium (223) | 88 Ra Radium (226) | 89-103 | 104 Rf Rutherfordium (267) | 105 Db Dubnium (268) | 106 Sg Seaborgium (269) | 107 Bh Bohrium (270) | 108 Hs Hassium (277) | 109 Mt Meitnerium (278) | 110 Ds Darmstadtium (281) | 111 Rg Roentgenium (282) | 112 Cn Copernicium (285) | 113 Nh Nihonium (286) | 114 Fl Flerovium (289) | 115 Mc Moscovium (290) | 116 Lv Livermorium (293) | 117 Ts Tennessine (294) | 118 Og Oganesson (294) |

| | | | | | | | | | | | | | | |
|---|---|---|---|---|---|---|---|---|---|---|---|---|---|---|
| 57 La Lanthanum 138.91 | 58 Ce Cerium 140.12 | 59 Pr Praseodymium 140.91 | 60 Nd Neodymium 144.24 | 61 Pm Promethium (145) | 62 Sm Samarium 150.36 | 63 Eu Europium 151.96 | 64 Gd Gadolinium 157.25 | 65 Tb Terbium 158.93 | 66 Dy Dysprosium 162.50 | 67 Ho Holmium 164.93 | 68 Er Erbium 167.26 | 69 Tm Thulium 168.93 | 70 Yb Ytterbium 173.05 | 71 Lu Lutetium 174.97 |
| 89 Ac Actinium (227) | 90 Th Thorium 232.04 | 91 Pa Protactinium 231.04 | 92 U Uranium 238.03 | 93 Np Neptunium (237) | 94 Pu Plutonium (244) | 95 Am Americium (243) | 96 Cm Curium (247) | 97 Bk Berkelium (247) | 98 Cf Californium (251) | 99 Es Einsteinium (252) | 100 Fm Fermium (257) | 101 Md Mendelevium (258) | 102 No Nobelium (259) | 103 Lr Lawrencium (266) |

- C Solid
- Hg Liquid
- H Gas
- Rf Unknown

# FOR FURTHER READING

Congdon, Lisa. *The Illustrated Encyclopedia of the Elements: The Powers, Uses, and Histories of Every Atom in the Universe*. San Fransico, CA: Chronicle Books, LLC, 2021.

DK. *Super Simple Chemistry: The Ultimate Bitesize Study Guide*. New York, NY: DK Publishing, 2020.

McHenry, Ellen Johnston. *The Chemical Elements Coloring & Activity Book: A Fun and Interactive Guide to the World of Chemistry*. Independently Published, 2021.

Silver, Donald, and Patricia Wynne. *My First Book About Chemistry*. Mineola, NY: Dover Publications, 2020.

Spangler, Steve. *Steve Spangler's Super-Cool Science Experiments for Kids: 50 Mind-Blowing STEM Projects You Can Do at Home*. New York, NY: Media Lab Books, 2021.

Thomas, Isabel, and Sara Gillingham. *Exploring the Elements: A Complete Guide to the Periodic Table*. New York, NY: Phaidon Press, Inc. 2021.

Woodbury, Rebecca. *Atoms and Molecules Meet*. Real Science-4-Kids, 2024.

Zovinka, Edward P., and Rose A. Clark. *A Kids' Guide to the Periodic Table: Everything You Need to Know About the Elements*. Emeryville, CA: Rockridge Press, 2020.

# INDEX

# ABOUT THE AUTHOR

**KATHLEEN A. KLATTE** is the author of many nonfiction books for children and teens. Topics she has written about range from animals and nature to constitutional law to unusual career choices. She lives in New York with one cat and far too many books and Legos.

# PHOTO CREDITS

Cover, p. 1 remotevfx.com/Shutterstock.com; Cover, pp. 1, 3-64 pluie_r/Shutterstock.com; Cover, p. 1, 3-64 Oksancia/Shutterstock.com; p. 5 marzia franceschini/Shutterstock.com; p. 6 foto_and_video/Shutterstock.com; p. 7 Dzuba Pavel/Shutterstock.com; pp. 8, 16 Peter Hermes Furian/Shutterstock.com; p. 9 albedomusic/Shutterstock.com; p. 10 Adriana Iacob/Shutterstock.com; p. 13 Adisak Riwkratok/Shutterstock.com; p. 14 Antoine2K/Shutterstock.com; p. 17 Tricky_Shark/Shutterstock.com; p. 19 RHJPhtotos/Shutterstock.com; p. 20 InWay/Shutterstock.com; p. 23 EJ Grubbs/Shutterstock.com; p. 24 Michael LaMonica/Shutterstock.com; p. 25 Selcuk Oner/Shutterstock.com; p. 26 Denis Belitsky/Shutterstock.com; p. 27 Chem_Geo_Photo/Shutterstock.com; p. 28 Vadim Sadovski/Shutterstock.com; p. 29 ImageBank4u/Shutterstock.com; p. 31 Nordroden/Shutterstock.com; p. 32 vyskoczilova/Shutterstock.com; p. 33 Nuttawut Uttamaharad/Shutterstock.com; pp. 35, 37 Tatjana Baibakova/Shutterstock.com; p. 39 Yankovsky88/Shutterstock.com; p. 40 Woodpond/Shutterstock.com; p. 41 DIAMOND HEART/Shutterstock.com; p. 43 Chadchai Krisadapong/Shutterstock.com; p. 44 Vladimir Zotov/Shutterstock.com; p. 45 Ground Picture/Shutterstock.com; p. 46 YARUNIV Studio/Shutterstock.com; p. 49 GraphicsRF.com/Shutterstock.com; p. 50 Alysson M/Shutterstock.com; p. 51 zhengzaishuru/Shutterstock.com; p. 52 Shutterstock AI/Shutterstock.com; p. 53 Yankovsky88/Shutterstock.com; p. 54 ShannonChocolate/Shutterstock.com; p. 55 View_From_My_Lens/Shutterstock.com.